How Animals Communicate

How Animals Communicate

by ANABEL DEAN

Drawings by Haris Petie

JULIAN MESSNER NEW YORK

Published by Julian Messner, a Simon & Schuster Division of Gulf & Western Corporation. Simon & Schuster Building, 1230 Avenue of the Americas, New York, N.Y. 10020.

Printed in the United States of America

Design by Haris Petie

Library of Congress Cataloging in Publication Data

Dean, Anabel.
How animals communicate.

Includes index.
SUMMARY: Discusses how and why animals communicate.
1. Animal communication – Juvenile literature.
[1. Animal communication] I. Petie, Haris. II. Title.
QL776.D4 591.5'9 77-23281
ISBN 0-671-32818-2

Contents

BOOKS BY ANABEL DEAN AND HARIS PETIE

How Animals Communicate

Animals that Fly

How Animals Communicate

CHAPTER ONE

What Is Communication?

When you talk with a friend, that is communication. But you do not have to talk to communicate. You can shake your head when you mean "no," put your finger to your lips when you mean "quiet." Your arm around a friend who has skinned a knee shows you are sorry. *Communication takes place any time a message is sent and received.*

From human beings down to the smallest insect, every animal has some way of communicating. For animals to survive as a *species* — a group with characteristics in common — they must have a food supply, a safe place to live, a mate and young. To get these things, they must communicate.

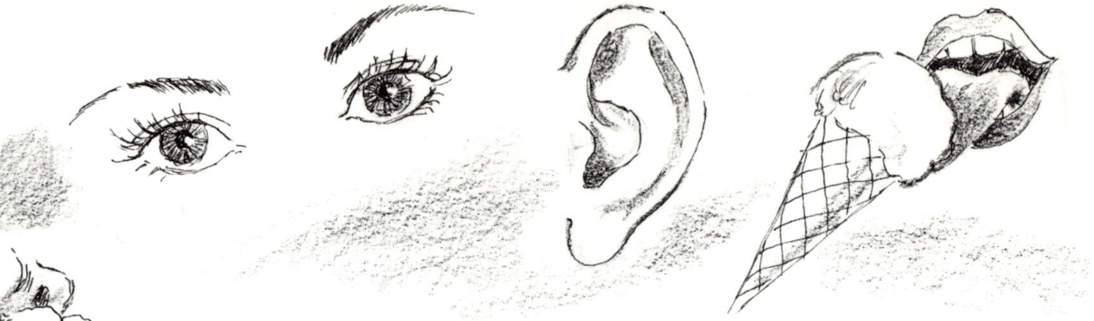

RECEIVING SIGNALS

Sensory organs — specialized parts of the body — receive communication signals. Eyes see, ears hear, the tongue tastes, the nose smells, and the skin feels. Tiny electrical signals from these sensory organs speed the information they receive to the brain. The brain interprets these signals: see whom? smell what?

In most animal groups, some of the sense organs are better developed than others. These, then, are the senses most used for communication by that animal.

Visual Signals. The eyes receive visual (sight) signals. Light-sensitive cells of the part of the eye known as the *retina* receive the pictures and send them on to the brain.

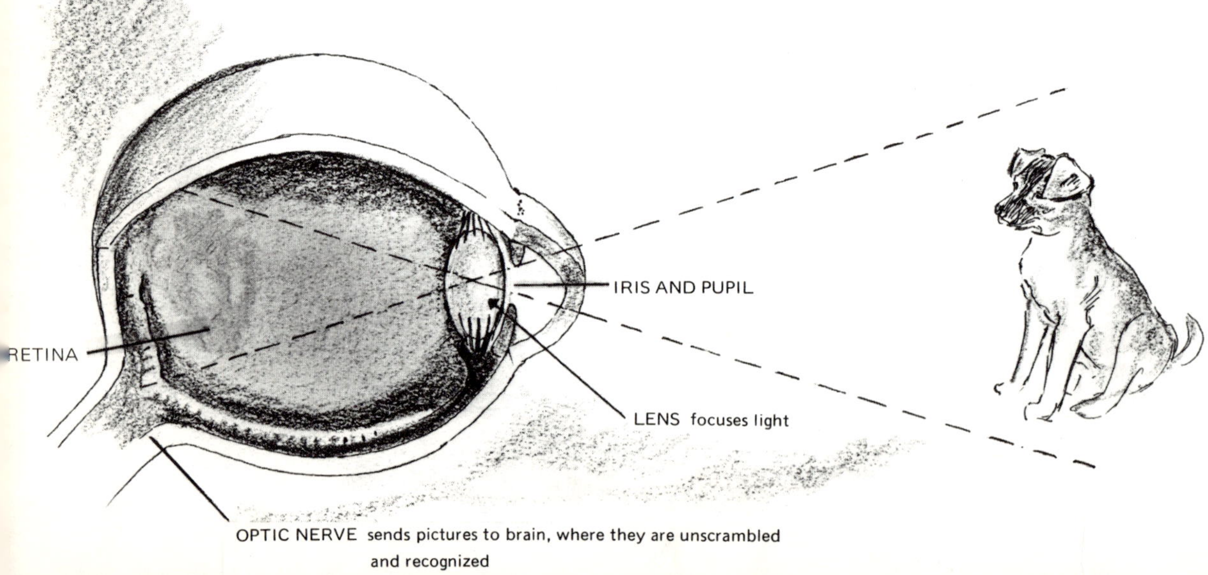

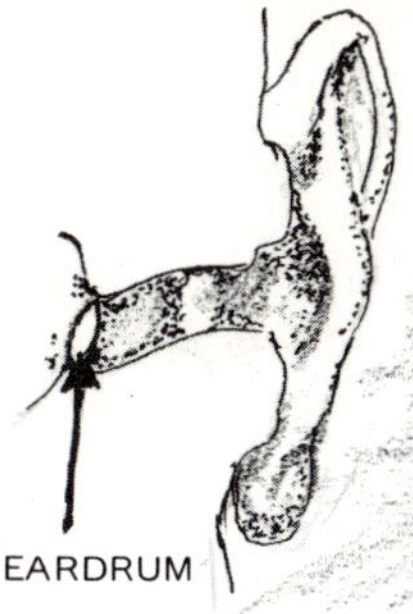

Sound Signals. Animals receive communication signals from sound waves carried through air or water. These waves strike a *membrane* — a thin kind of skin. In human beings and some animals, this membrane is called an *eardrum*. The sound waves make the membrane vibrate, and the vibrations are sent to the brain. Here they are recognized as sounds.

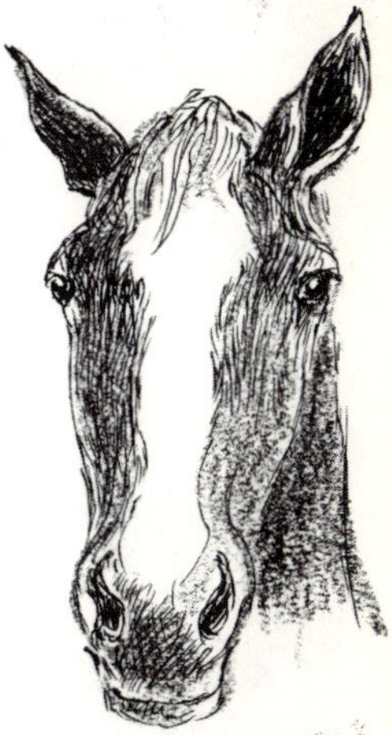

Human beings have outside ears to funnel sound into the eardrum. All *mammals* — humans and other animals whose young are born alive — have ears. Some, such as horses, dogs, and deer, can turn the outside ear to catch sounds.

Birds have eardrums, but no outside ear.

Frogs, toads, and lizards have membranes on the sides of their heads which pick up sound waves.

Insects usually have a hearing membrane located on their legs or abdomens. Some pick up sound waves with sensitive cells in their *antennae* — long, slender, hair-like, sensory organs.

Snakes, fish, and turtles do not have any kind of ear openings. Their skins pick up sound vibrations coming through the water or ground.

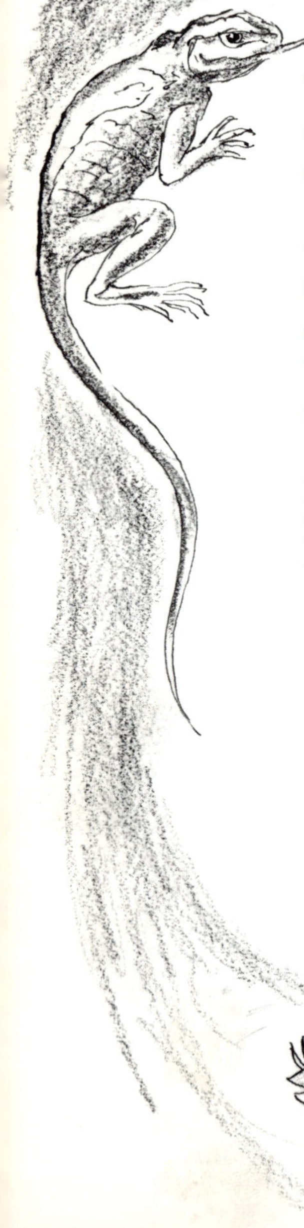

Taste and Smell Signals. The nerves carrying taste and smell signals to the brain are closely linked. That's why, when our noses are clogged, food doesn't have much taste.

In animals with backbones, *vertebrates*, the cells picking up the chemical particles that carry odors are inside the nose. Snakes and lizards also have small openings inside their heads, called Jacobson's organs, to gather smell and taste signals. Their tongues flick in and out, picking up and carrying the chemicals of smell and taste to these organs.

The sense of smell in insects is located on the antennae. In fish it is located in two small pits on the snout. Some fish have taste buds on their fins, others have them on the head. The catfish has them in its whiskers.

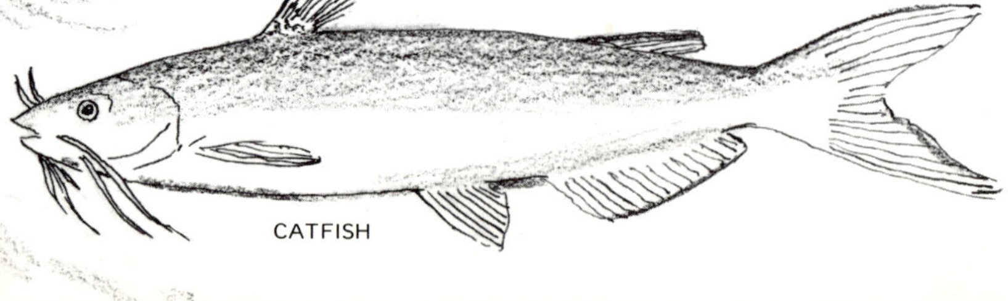

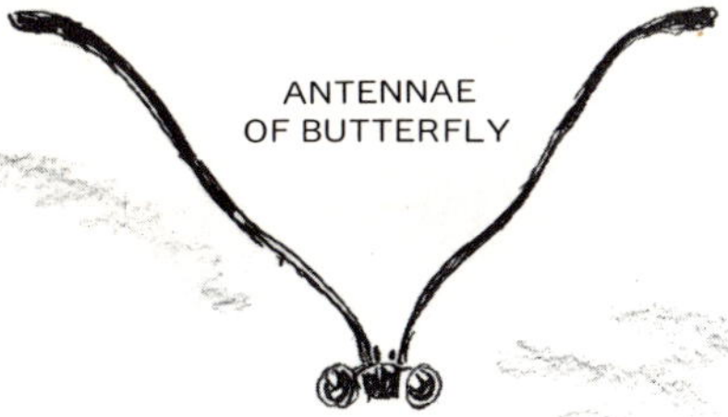

Touch Signals. In many animals, the sense of touch is carried to the brain by nerve endings just under the surface of the skin. Some insects have nerve endings in tiny hairs on their bodies. In other insects, the antennae or feelers are special touch organs. Some fish also use feelers for communicating by touch.

SENDING SIGNALS

Sound Signals. Animals use sound signals to send messages over a distance. These sound signals are made in different ways. Mammals, and some reptiles, have a voice box in the throat called a *larynx*. Two *vocal cords* — bands of tissue — stretch across this voice box. Air from the lungs is passed over the vocal cords, causing them to vibrate, making waves of sound which are carried through air or water.

Frogs and toads have *vocal sacs* as well as vocal cords. Air can be stored in the vocal sacs and forced back and forth between the lung and the vocal sacs.

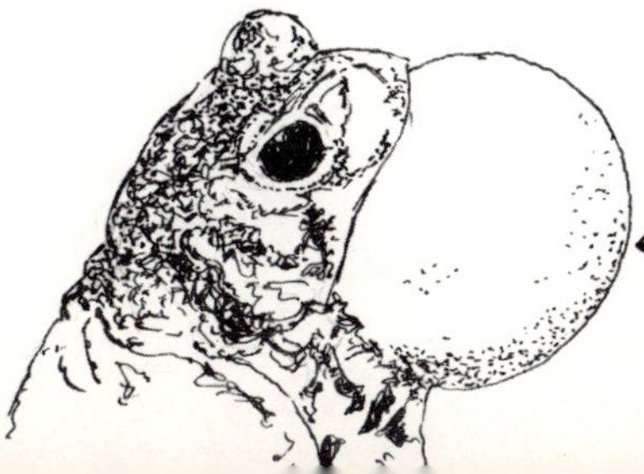

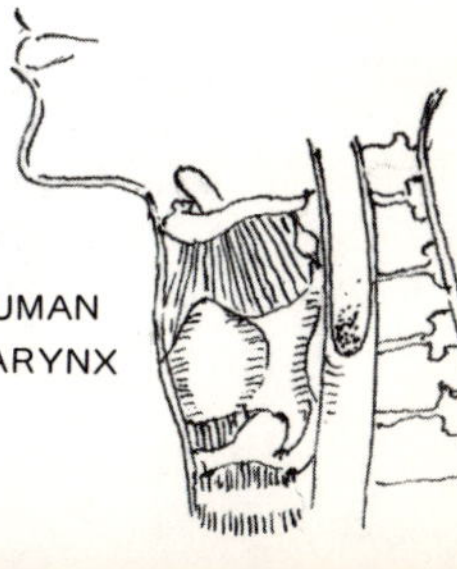

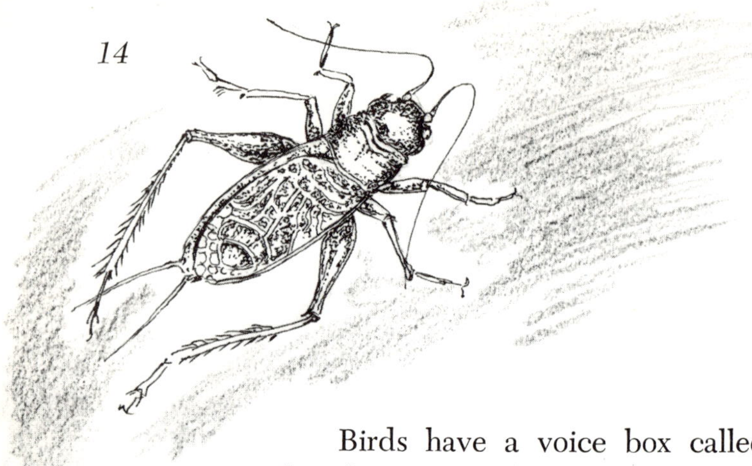

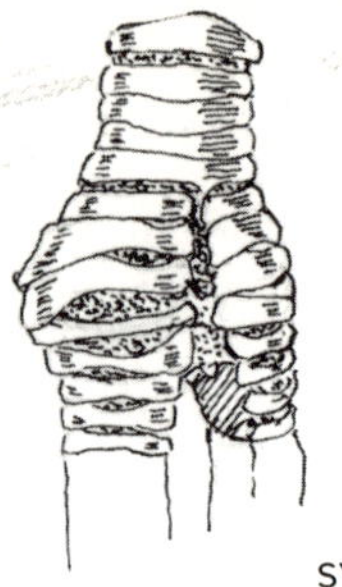

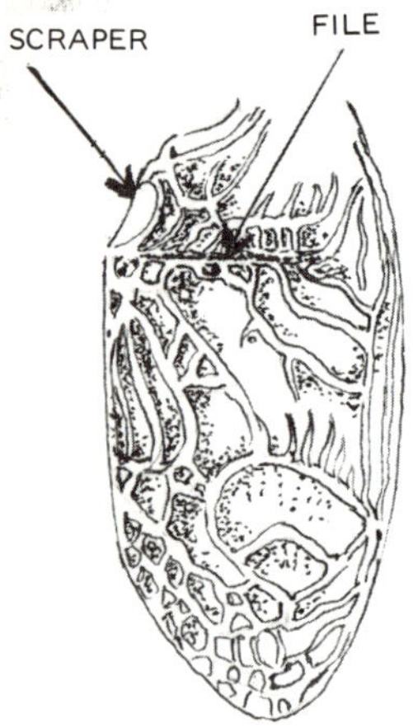

Birds have a voice box called a *syrinx*. Sound is produced as air is forced from the lungs through membranes covering the two openings in the syrinx.

Insects do not have voice boxes. Most make sounds by rubbing one part of their hard bodies against another.

Fish do not have voice boxes either, but make sounds in other ways. Some fish produce a drumming sound by vibrating the walls of their air bladder. Some fish beat their gill covers, and some grate their teeth together.

There are many other ways of communicating with sounds that are not vocal. A beaver slaps the water with its tail as an alarm signal. A porcupine rattles its quills and a gorilla beats its chest as a warning message.

Taste and Smell Signals. Some animals communicate with chemicals called *pheromones* (fer-uh-moanz). Fish release pheromones in the water. Some insects and snakes lay these chemical trails on the ground. Moths release pheromones in the air. These smells may carry an alarm or warning. They may be used to attract a mate, or to tell other animals that a territory has been taken.

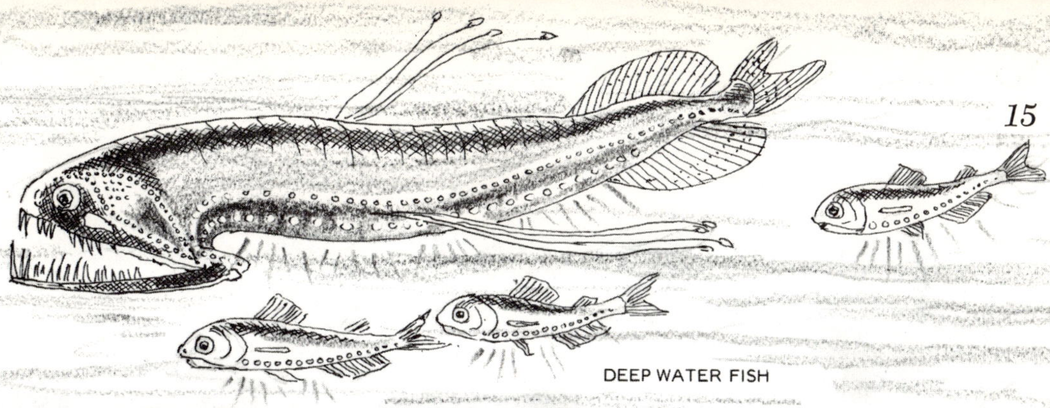

DEEP WATER FISH

Visual Signals. Many animals send messages that must be received by the eyes.

Some birds show off their plumage (feathers) as a signal, particularly if their plumage is beautiful and their song is not. Others do a little dance. These signals are called *display*, and are usually used to attract a mate.

Some deep water fish flash lights as signals. Members of the monkey and ape families use facial expressions much like ours to communicate anger, love, and other emotions.

JELLYFISH

Touch Signals. Most animals can send out touch signals with the same organs they use to receive them. The antennae of some insects, the feelers of some kinds of fish, and the arms of an octopus are all used this way.

Some animals, however, may receive touch on one part of their body, and send out touch with some other part. Humans do this. If you touch a person too hard on any part of the body, that person may use his or her hands or feet to touch you back, very hard!

If you touch a jellyfish on any part of its body, it will try to sting you with its long slender arms, or tentacles — its organs of touch.

CHAPTER TWO

Everyone Communicates

People Communicate. What would you do if you were to bang your head on a branch while traveling through the jungles of Zambia? You would probably say "Ow!" and rub your head. If a Zambian hit his head on the same branch, he would do the same thing. A cry of fear, joy, or pain is the same no matter what the language. Gestures, also, often mean the same thing to many different peoples. People communicated with cries and gestures long before they developed languages.

But cries and gestures do not give exact information. It would be hard to grunt an order for a super-duper sundae with three scoops of chocolate fudge ice cream, strawberries, marshmallow sauce and nuts, and have it come out right!

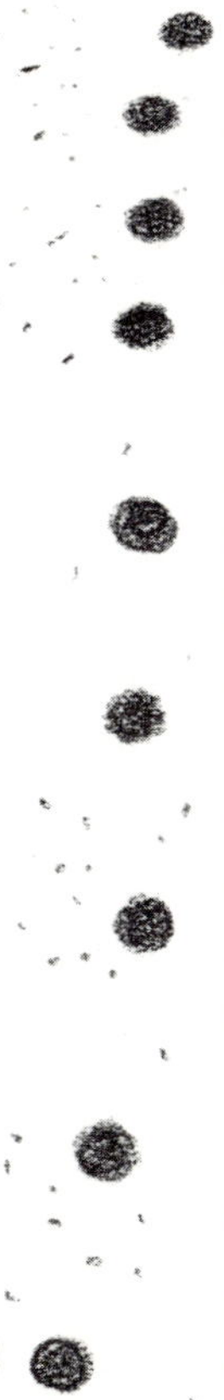

Words probably came into use one by one. One small group of people might have started to use a certain sound for some object — let's say a rock. After a long time, everyone from that tribe used the same sound for rock. It took thousands of years for a spoken language to develop.

For many centuries, spoken language was the only way to pass on knowledge. Over distances and years, much information was lost forever. Then people began to keep records by making marks in the dirt or scratching and painting pictures on cave walls or rocks, and, much later, on clay tablets. It took hundreds of years more for groups of people to develop an alphabet, with a letter to represent each sound they used. Then the letters could be put together to form any words. This is the method we use with our alphabet of 26 letters.

Writing made it possible to communicate over distances, and time. We read books written by people who died long before we were born. We communicate with letters and newspapers and radio and television pictures. Pictures and words can be sent around the world and to and from the moon — and Mars.

Humans and Animals Communicate. People and animals communicate by learning each other's signals. The hunter's scent carried downwind warns the deer to run. The shepherd directs his sheep dog with a motion of his arm. Many dogs have been able to follow the scent trail left by a human after smelling an article of wearing apparel.

A dog brings a ball and lays it at someone's feet to show it wants to play. We snap our fingers for the dog, or whistle. We call, "Here, kitty, kitty," and the cat comes to us. Other types of sounds may also carry messages. The sound of the electric can opener may bring the dog or cat tearing into the kitchen. The jingle of the car keys may send the dog running to the door.

People also receive messages from the sounds animals make. A growling dog or hissing cat will make a person retreat. The purr of a cat or wagging tail of a dog give people a different message.

Messages are passed between animals and human by touch. A cat rubs against a person's legs when it wants something. You pat your dog to show approval.

Experiments are being made to see if an animal can be taught to communicate with a human language. Keith and Cathy Hayes took Viki, a six-week-old chimpanzee, and raised her as if she were a child. In six years, Viki learned many directions, so she understood many words. But speaking them was too difficult — she only learned to say three words.

In 1966 Allan and Beatrice Gardner took a chimpanzee, Washoe, into their home to try to teach her the sign language used by the deaf. When Washoe was four, she could use more than sixty words in sign language. She made the sign for "please" by drawing an open hand across her chest. She could say "I want more" by putting her fingertips together.

After Washoe learned the word for "open" when she wanted the door to her room opened, she was able to understand that she could use the word for anything she wanted opened. She also learned to put words together to make sentences.

Trained dolphins and whales can learn verbal directions. They also try to imitate human speech. Their imitations are sometimes used as signals. But it is doubtful that dolphins can be taught human speech, since they do not have vocal cords.

In Hawaii, a scientist has reduced the Hawaiian language into whistles. Dolphins have learned to obey the whistle commands and even answer with the same whistles.

Will humans and any of the animals ever be able to talk together using human language? The answer to that question still remains to be answered.

Animals Communicate. Animals communicate in an amazing variety of ways. But they must see, hear, smell, or touch each other to exchange messages.

The information animals exchange can be complex. One bee can direct others to the exact location of a source of nectar several miles away. Two coyotes, hunting together, often work out a plan beforehand to trick their prey.

Unlike humans, many animals are born knowing their communication signals, or most of them. In one experiment, a baby cheetah was taken from its mother as soon as it was born. Although raised alone, it still used all the signals cheetahs use. But other animals must be taught some of their signals by others of their species.

Let's find out what the animals are communicating — and how they do it.

CHAPTER THREE

Communicating To Live

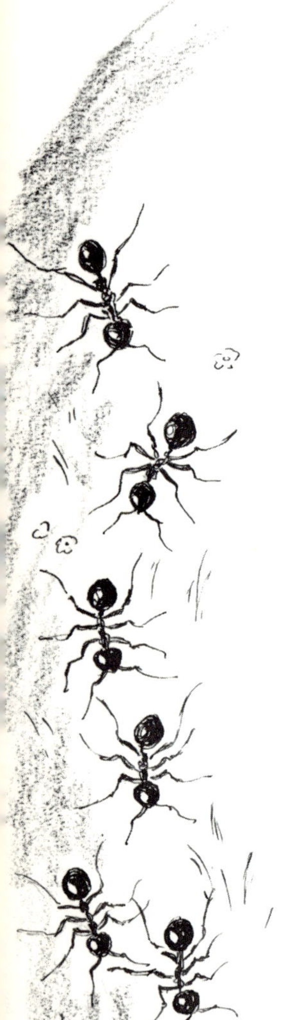

FINDING FOOD

All animals need food. Social animals — those living in groups — often work together to get food and share it. Signals are used so each knows what to do.

Smell Signals. Have you noticed how quickly ants find food at a picnic? First you see only one or two ants. These are scouts, searching. Soon there is a trail of ants. When the scouts find food, they hurry back to the nest, the smell glands on their abdomens leaving a chemical trail. Other ants then follow the scent trail to get the food. As they return to the nest, they, too, leave scent on the trail so it will not fade. But as the supply of food gets smaller, less scent is left. Fewer ants go out to get food. When the food is gone the last ant leaves no scent trail on their way home.

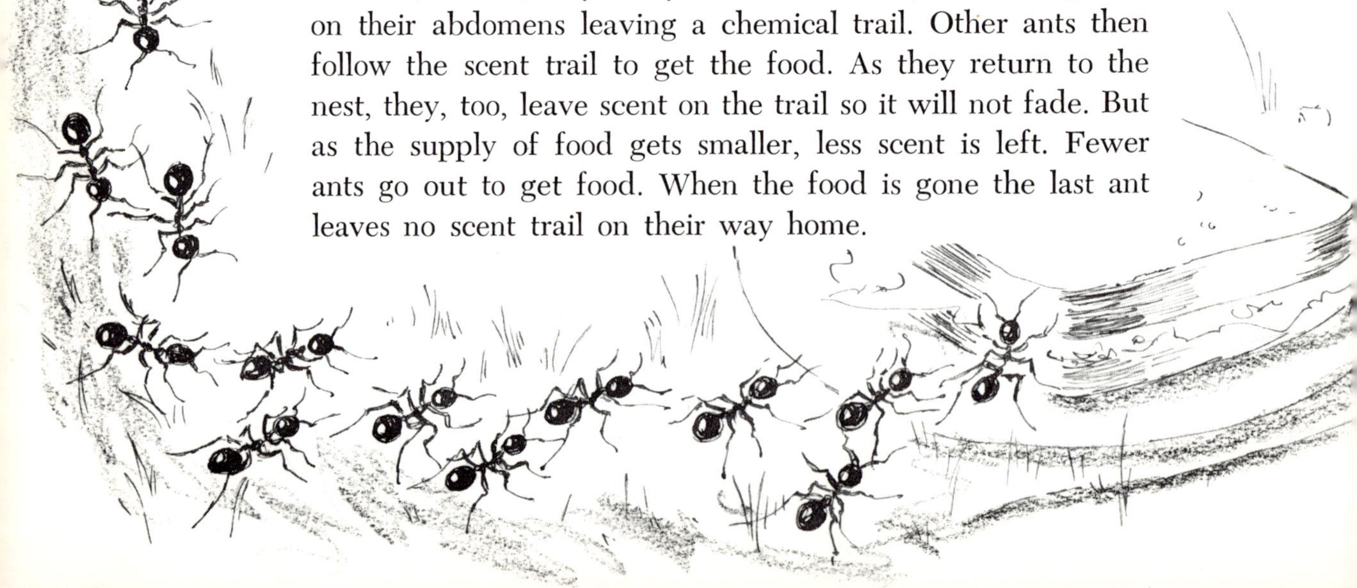

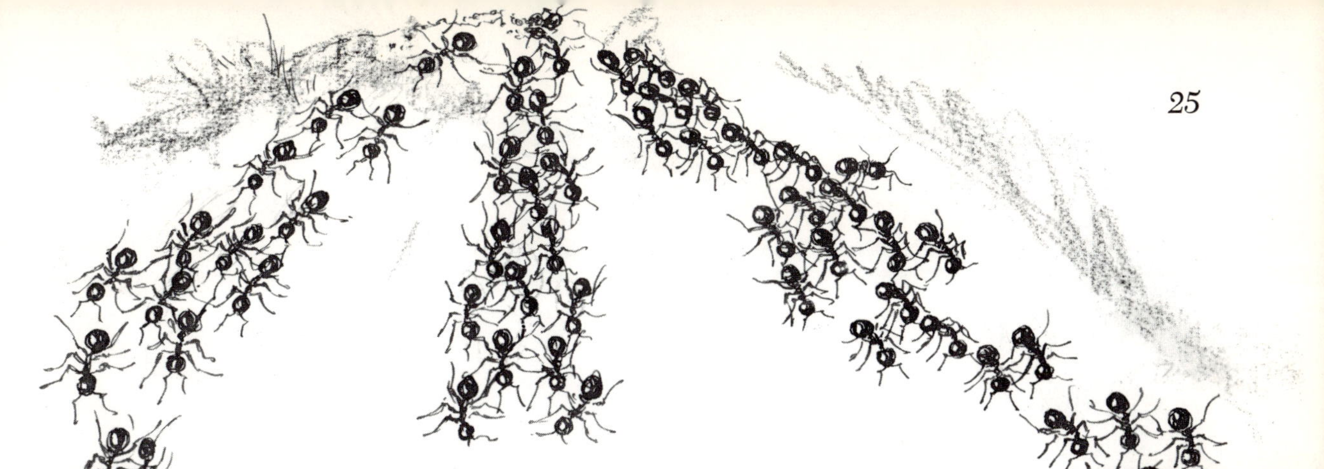

One scientist tried an experiment to see if ants could tell others how much food there was to bring in. He put three different size pieces of grasshopper down near an ant colony. The second piece of grasshopper was twice the size of the first. The third piece was twice the size of the second. Twenty-eight ants went out to get the first piece of grasshopper, 44 went out to get the second piece and 89 went out to get the third piece.

Visual Signals. Some animals use visual signals when they find food. When one crab finds food and starts to eat, others scurry toward it. The crabs farther away see these crabs leave and they, too, hurry over.

The same system is used by vultures. Vultures are widely spaced as they circle far above the ground looking for food. But they can see each other. When one vulture goes down to investigate something, those closest follow it. Vultures still farther away see these others leave, and they, too, go down to investigate. Within a short time many vultures have gathered around a dead animal on the ground.

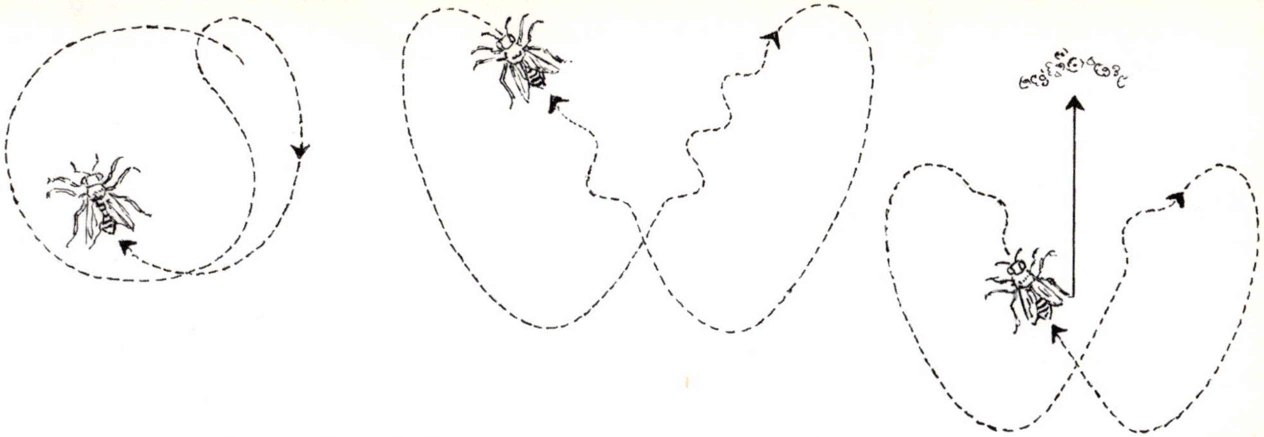

When a bee finds nectar, it returns to the hive to tell others about it with a dance. The bee wriggles her tail back and forth as she moves straight ahead, then circles to the right. Then she moves straight ahead again, and circles to the left. She does this dance over and over. The bee usually dances on a honeycomb which is standing on end in the hive. If the bee does the straight part of the dance *up* the comb, the food is located in the direction of the sun's position in the sky. If the straight part of the dance is done *down* the comb, the food is located away from the sun. If the bee does the straight part of the dance 45° to the right of the top of the comb, the food is located 45° right of the direction of the sun, and so on. This wriggling dance is repeated quickly when food is near the hive. The farther the food is from the hive, the slower the dance.

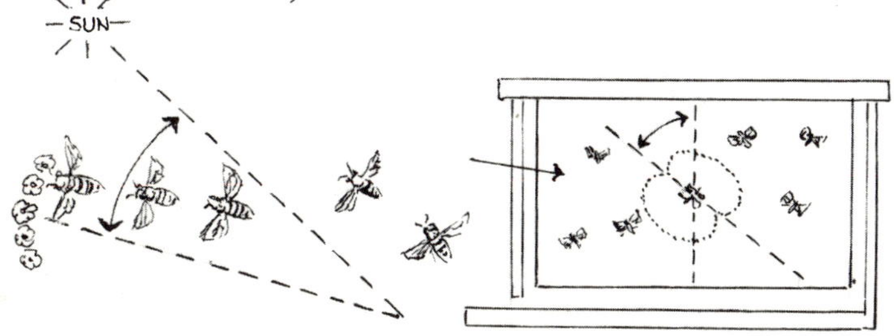

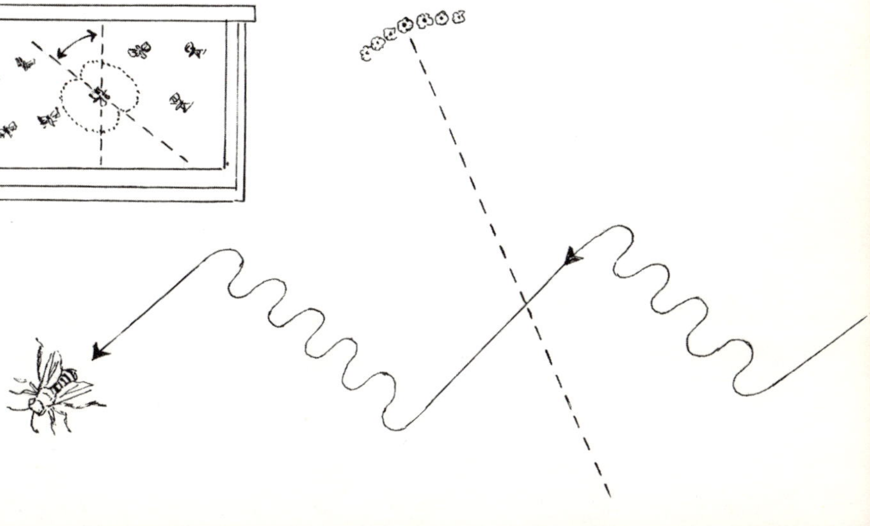

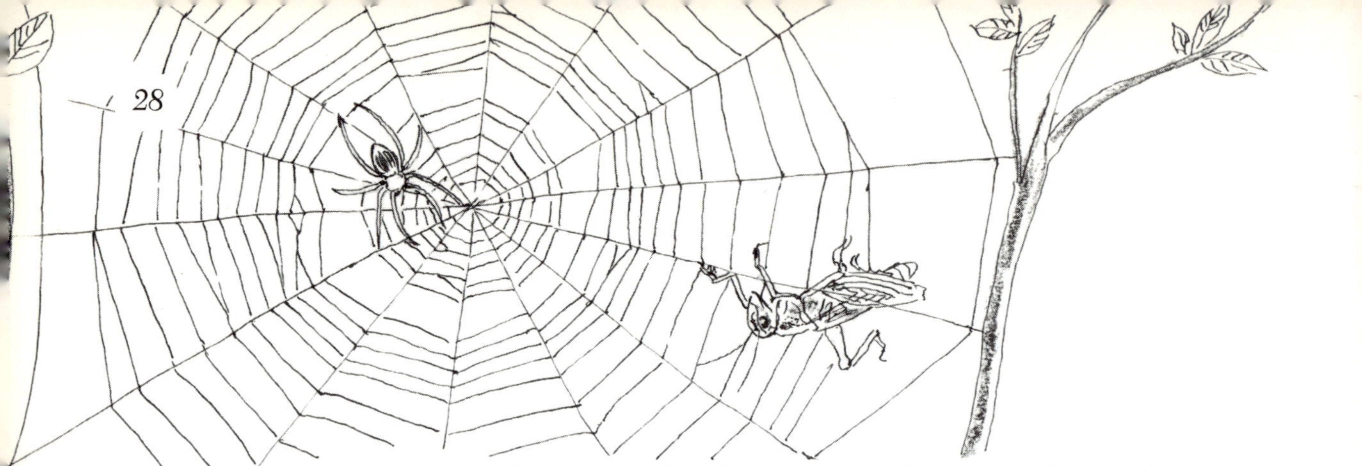

Touch Signals. Some female spiders build webs. They wait with one foot on the spoke of the web. When an insect is caught, the spider can tell from the vibrations the location and size of the catch.

Sound Signals. Many birds give feeding calls. When the hen scratches up a worm, her *cut-cut-cut* calls her chicks.

If a gull finds a large amount of food, it gives a loud food call which brings gulls from all over. But if it finds only a little food, the gull gulps it down quickly without making a sound.

Young birds give a begging call when they are hungry. At first, they squawk all the time with wide-open beaks. Later, they call only when a parent alights on the edge of the nest.

When animals hunt together, they use signals so each knows what to do. Wolves communicate with barks when the pack hunts. Lions use a low cough. Animals raised alone can't hunt in a group, because they don't know the signals.

Chimpanzees make some sounds which always mean a certain thing, just as our words do. To a champanzee *kuoh* always means "I want food." *Gak* means "I have found food." It may be that chimpanzees are slowly developing their own language just as human beings once did.

A PLACE TO LIVE—TERRITORY

Your home belongs to you and your family. Others can't come there unless you want them to. Animals, too, must have a place of their own — a territory. If animals live in a group, the territory belongs to all. The territory is just large enough to provide the group with food and a place to live. No others of the same species are allowed there. Communication signals are used so animals know the boundaries of each other's territory, and so avoid fights.

Smell Signals. Social insects recognize members of their own colony by odor. When an ant arrives at the nest, the guards touch antennae with it. If this ant doesn't give off the right smell, the guards kill it. Bees do the same thing. Sometimes a bee alights at the wrong hive by mistake. The guards rush up to check it. When they find it doesn't have the right smell, they pull its wings and legs off and push it off the landing board.

Male animals leave smell signals around their territory. Others of that species pay attention to these "keep out" signals unless they want a fight.

Rabbits produce a colorless fluid from glands under their chins. This is rubbed on the ground and plants around their territory.

Beavers mark out their territory with "beaver pies" made of mud mixed with chips of wood, sticks and pebbles. A few drops of strong musk is added so the small pancake-shaped "pies" can be smelled as well as seen.

Bears leave scent around their territory by plastering mud on trees and then rubbing their backs against the mud. Hairs rub off, carrying the bear's smell with them.

Visual Signals. Some animals use visual signals to warn others away from their territory. The common fence lizards, sometimes called blue bellies, signal to each other with the blue markings on their bellies and throats. Each male stands on a high spot such as a fence or rock. Then he uses his legs to push his body up and down showing his blue belly in the sunlight.

Male bears mark out their territory by reaching up as far as possible and scratching tree trunks. Other males look for these scratches. The marks tell them not only the location of another bear's territory but how big the bear is. If the new bear is larger, he often puts his scratch marks above the other's. The black bear always leaves four scratches, the grizzly five.

Sound Signals. Male birds usually arrive at the nesting sites before the females. Each picks out a territory and sings to warn other males away. If another male gets too close, there is a singing match — usually won by the bird who got there first.

Some birds use non-vocal sounds for declaring territory. The male ruffed grouse perches on a hollow log and makes a drumming noise by rapidly fanning its wings. The male woodpecker taps on a dead tree.

Prairie dogs live in large groups, known as "towns." Although they use the same feeding ground, each male watches over his own burrow. When he sees an intruder, he stands on his hind legs and gives a loud whistle. This means, "this is my territory — keep out."

CHAPTER FOUR

Communicating To Raise A Family

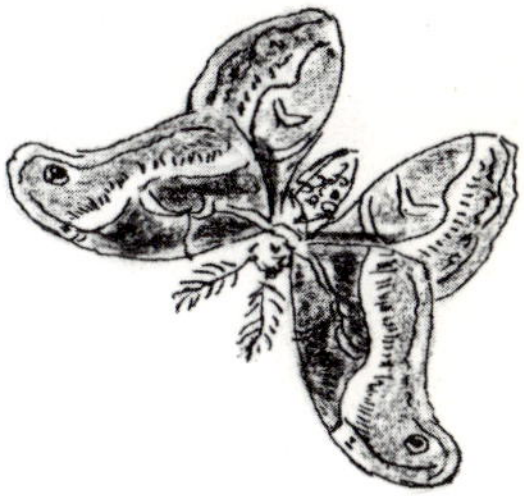

FINDING A MATE

Some animals live alone, or the males live apart from the females. For each animal to find a mate at the right time, courtship signals are used. Since animals mate only with those of their own species, signals are exchanged to be sure they recognize one another.

Smell Signals. Scent is often used as a mating signal. The scent of a female moth attracts males from over a mile away.

Female weasels leave a scent trail when they are ready to mate. If a male weasel catches the scent, he follows, and leaves a scent trail of his own to discourage other males.

Snakes also leave a scent trail which tells their sex and species.

Bull alligators release a strong scent from glands under the jaws to attract females. This odor hangs over alligator swamps in the spring.

There are sometimes only a few fish of one species in miles of water. The females will release a chemical a few weeks before they are ready to lay eggs. This gives males the time needed to catch the scent and find them.

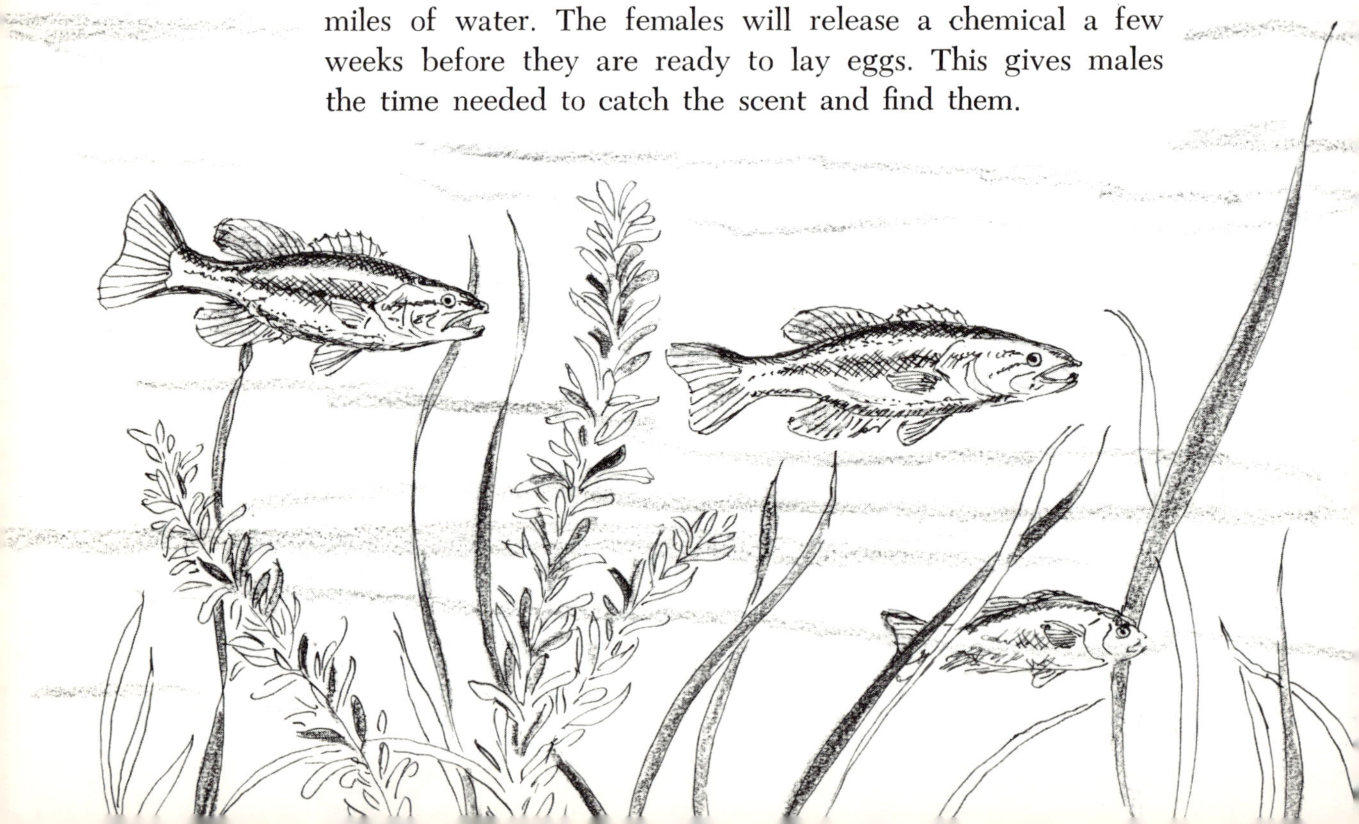

Visual Signals. Some sea animals change color or pattern of colors as a signal that they are looking for a mate. The male of one species of octopus usually has stripes running across his body. But when courting, the stripes run up and down. Both the male and female of one species of cuttlefish show zebra stripes when courting.

Many male crabs wave their brightly colored large claw to attract a mate.

Far down in the ocean it is very dark. Some of the animals living there send messages with light. One marine worm gives off a flashing light when looking for a mate. And, as the female worms only give off a slight glow, many of them gather together into a glowing ball to be more noticeable.

Fireflies also use light signals to attract a mate. On a spring evening the males fly close to the ground, flashing a light on the end of their abdomens. The females remain in the grass, answering with a paler light.

Birds sometimes use courtship displays to attract a mate. The peacock makes a fan of his long blue, green, and gold tail spotted with "eyes," and parades in front of the peahen.

When the feathers of male and female birds are the same, other visual signals are used to find a mate of the right sex. A male Adélie penguin picks up a pebble and places it in front of another penguin. If the second penguin is a male, he attacks the first penguin. If it is a female but not interested, she ignores the pebble. If it is a female who wouldn't mind being courted, she picks up the pebble.

All common terns look alike with their black caps and orange-red bills. When courting, a male tern catches a fish and walks up and down the beach with it. If a female is interested, she comes up and takes it from him. Often the two fly through the air, passing the fish back and forth.

The male and female roadrunner also look alike. When looking for a mate, the male catches a lizard and runs around with it dangling from his beak. If a female is interested she crouches down and flutters her wings. Then he gives her the lizard to eat.

The rather drab male bowerbird builds and decorates a bower — a little shelter — to tell the females he wants a mate. He first clears off a piece of dirt and gathers sticks to set in the ground. Some species make an enclosure of the sticks.

Others make a little tunnel. Then the bowerbird decorates his bower with berries, flowers, insect's wings, snail shells and feathers. He paints it with juice from berries crushed in his beak. Pieces of broken glass or the shiny top of a tin can are sometimes added. To attract a female into his bower, the male dances around in front of it picking up different ornaments in his beak to show her.

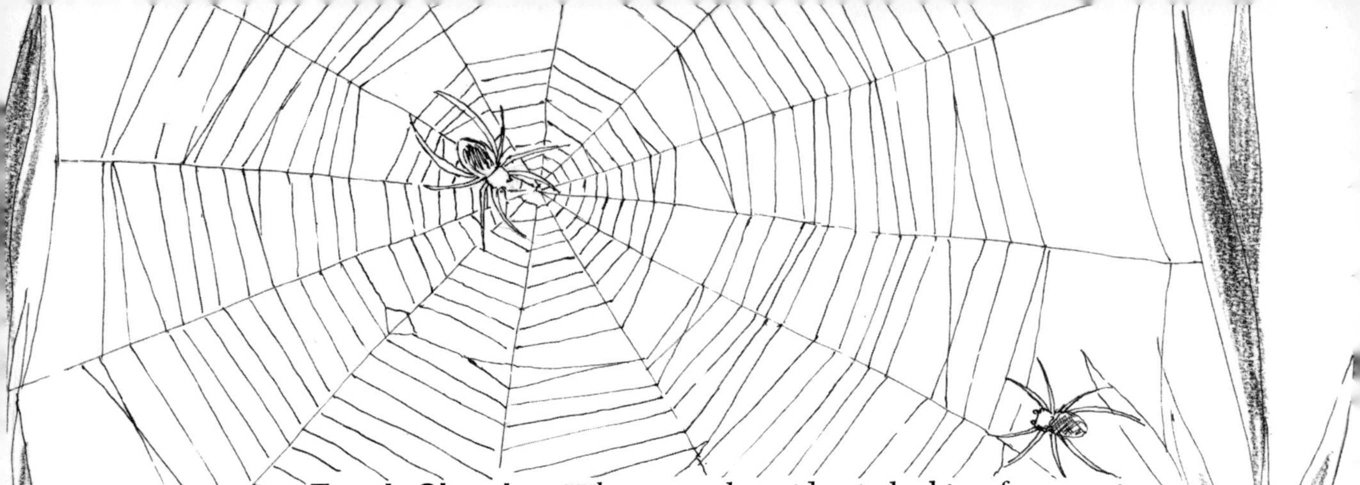

Touch Signals. When a male spider is looking for a mate, he comes to a female's web and plucks on one of the threads. If interested, she gently plucks the thread in answer. If not interested, she shakes the web. Then the male hurries away, as female spiders often eat the males.

Sound Signals. In spring or early summer, male frogs and toads gather at ponds and call to attract the females. Each species has a different call, ranging from the high-pitched "peep" of the spring peeper to the deep "chug-o-rum" of the bullfrog.

Usually, male insects give a mating call and the females move toward them soundlessly. In some species though, the female answers. Guided by her answers the male finds the female. Each species of grasshopper has a different call and is only answered by one of the same species.

Some of the larger mammals use sound signals to attract a mate. Bull elks make the mountains ring with their bugling calls in the fall. Only the older bulls give the full call. The younger ones are called "squealers" because of their shrill calls.

RAISING AND TRAINING THE YOUNG

Most reptiles, fish, insects, frogs and toads lay their eggs and forget them. Communication signals are not used between parent and young. But baby birds and mammals are helpless. Communication signals are used as the parents feed, care for, and train their young.

Visual Signals. Many parents teach their young to hunt by showing them how to do it. A lioness begins teaching her

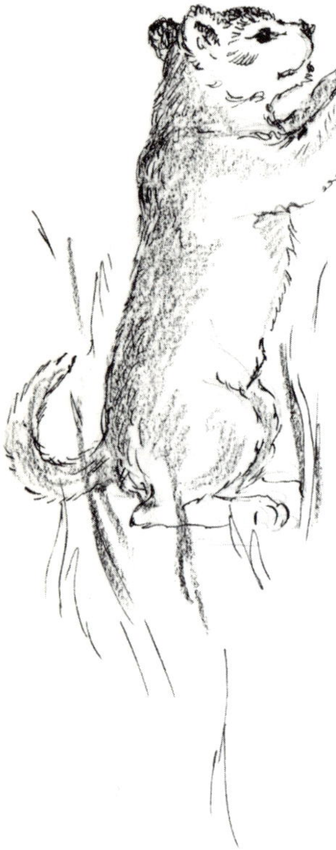

cubs when they are about five months old. First, she has them try to catch the tassel (tuft of hair) at the end of her tail as she moves it back and forth. Later, imitating the movements of the mother the cubs practice catching butterflies and other insects — then mice and other small game.

Wolves also work together to teach their young how to hunt in a pack.

Mother raccoon shows her young how to fish.

Young beavers are taught how to fell trees, build dams and houses, and get food. Beavers raised alone cannot build dams.

Touch Signals. When a young dolphin is hungry, it gives a signal by diving down and nuzzling against its mother's side. Then she *quickly* squirts milk into its mouth, because her baby must surface for air twice every minute.

Sound Signals. The mother alligator lays her eggs in a pile of decayed vegetation. When the young hatch out, they begin squeaking. Then the mother uses her long snout to dig them out. She leads them to water, giving low grunts to warn them of danger. When frightened, the young give little distress cries.

Most mammals use sound to bring mother to them when they are hungry or frightened. Young dogs, wolves and weasels whimper and whine. Baby rabbits and possums cry, kittens mew.

Each mammal mother has signals to tell her young to be quiet or hide when she senses danger. When the vixen — female fox — warns her kits with a sharp bark, they dive into the den. Wolf puppies are quieted down by a low-pitched growl. The weasel makes a little cooing noise to keep her young close as they search for food. A sharp snarl sends them into hiding. A mother skunk chatters to her young to keep them near. Mother bears give a low growl which means 'climb a tree'.

CHAPTER FIVE

Communicating To Live With Others

A PLACE IN THE GROUP

Most animals live in groups of two or more. Communication signals are used so each animal knows its place and that of others in the group. When the group travels, signals are used to keep everyone together.

Visual Signals. The dominant male — the leader — of a baboon troop walks stiff-legged with tail raised to show his authority. All others give way to him. If another baboon does something he doesn't like, the leader stares at it. But no one stares at a dominant animal. This is a threat. Scientists studying apes and monkeys in the wild learned not to look directly at them.

Members of the monkey family communicate with other facial expressions also. When one member of the group makes a noise, others look to see its expression. A scowl shows a fight is in the making. Affection is shown by pursing the lips and even touching them in a kiss.

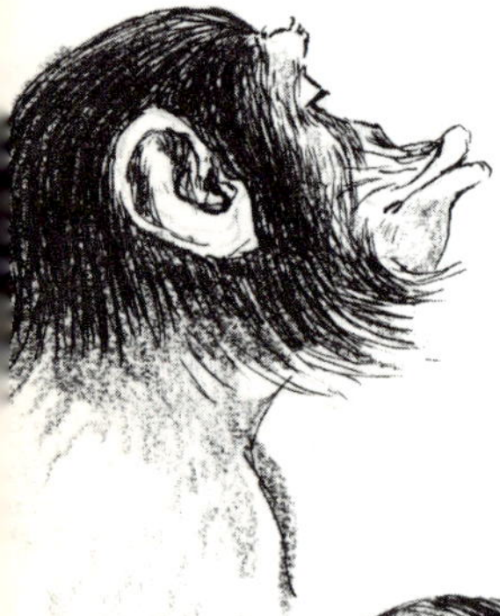

When one chimpanzee has food, the others gather around it, holding out their hands in a begging gesture. Chimpanzees will use an outstretched arm and finger to point out danger. They raise one hand in the "stop" gesture used by traffic officers or motion "come here" with their arm. A "stay back" gesture is sometimes used by a leader who wants to look around before the group goes out into a clearing.

Touch Signals. Mammals often show affection by touch—usually *grooming*. One animal combs the fur and picks out dirt for another. The one being groomed often closes its eyes with pleasure. After two friendly prairie dogs meet, one rolls over while the other grooms it. Then they trade places and the other is groomed.

Horses, cattle and deer show affection by nuzzling, licking, and neck rubbing. Elephants touch each other with their trunks and lower lips. Members of the cat family lick each other with their sandpaper-like tongues.

Chimpanzees, gibbons and baboons sometimes embrace each other upon meeting. Two chimpanzees often put out their hands and touch them in a "handshake." They also touch the other animal on the head or shoulder upon meeting. A lower ranking male cannot remain near the dominant male without permission. He asks to stay by holding out his hand, palm up. If the dominant male reaches out and touches the extended hand, the lower ranking animal may stay.

Sound Signals. Animals often use sound signals to keep the group together. Chickadees call to each other as they feed. Then one bird hops higher in the bushes, giving the flight call. The others answer and the entire flock takes off together.

When geese prepare for take-off, one bird gives the flight call, softly at first. Others echo the call, and gradually they all honk louder. A few geese begin to shake their heads in the departure signal. Then the entire flock skims over the water and takes off together.

Migrating animals — those moving to another area for the season — use signals to keep together. Sometimes we can hear the lonesome calls of ducks or geese as they pass high overhead. Migrating seals use short barks.

As their eyesight is poor, a herd of wild pigs grunt all the time as they look for food in the underbrush. When one pig strays away, it snorts. Then the others grunt a little louder to guide it back. Baby pigs learn these signals within a few days.

PROTECTING EACH OTHER AND THEMSELVES

For safety, groups don't allow strange animals in their territory. They give warnings when one comes too close. When a group member returns to its territory, it gives signals so the others will know it belongs there.

If one animal senses danger, it gives an alarm signal. An animal which has been hurt or is in danger gives a distress call. Others from the group sometimes try to help it.

Smell Signals. When a fish is hurt, it gives off a chemical which communicates alarm to other fish. If an injured minnow is placed in a school of minnows, all flee in alarm. When a little water in which an injured minnow has been swimming is poured into a school of minnows, they too show alarm and huddle at the bottom of the tank.

Other animals often try to raid harvester ants' nests to get the seeds they store. When a raiding party attacks, the ants at the danger spot release an alarm scent. Only the ants near the spot smell the scent and rush to attack the invaders. If the entire nest turned out for each alarm, no work would get done.

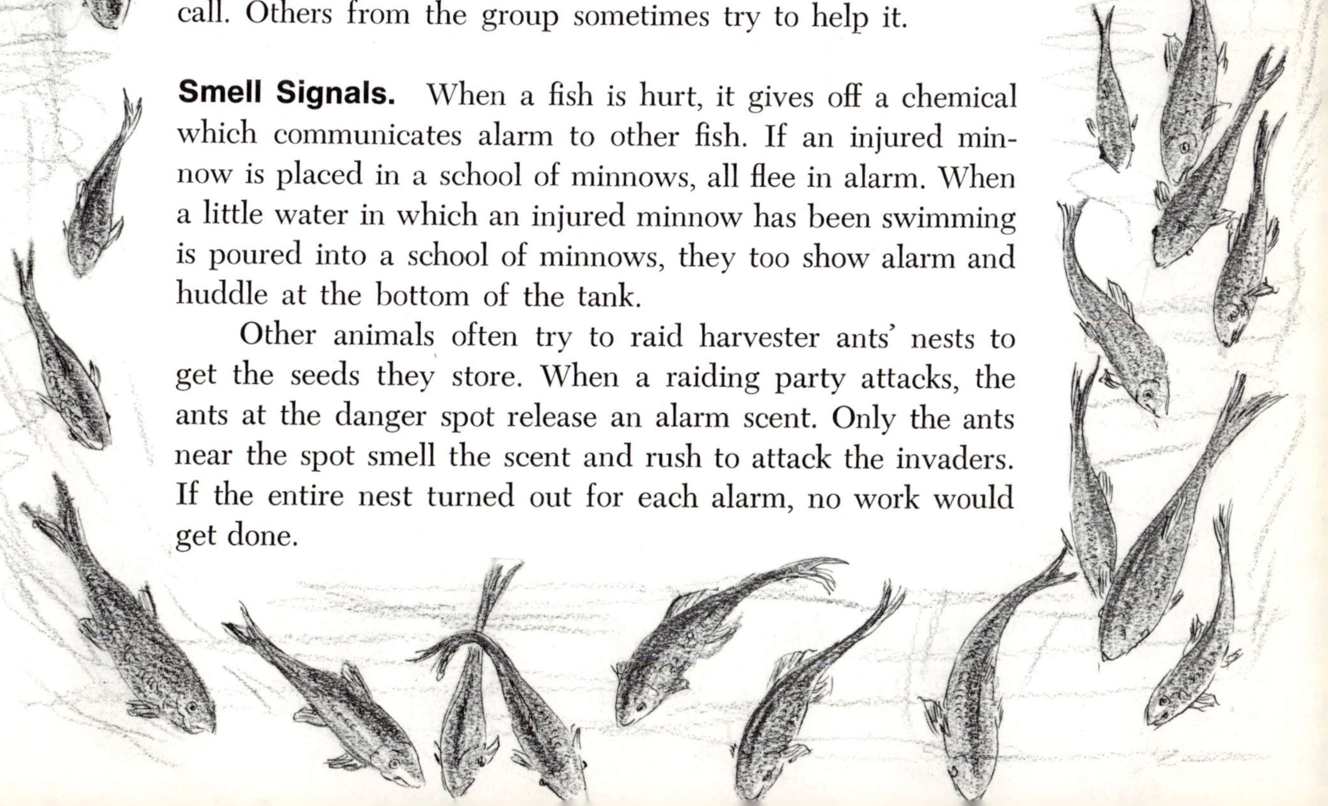

Visual Signals. Animals living in herds often flash alarm signals which can be seen at a distance. The oldest doe usually leads a band of white-tail deer. When she senses danger, she flashes the white underside of her tail as an alarm signal to the others in the herd.

Pronghorn antelopes use their keen eyesight to spot danger at a distance. They raise the white hair on their rumps as an alarm signal. This area becomes twice as large and can be seen by other antelopes up to two miles away.

Most animals don't want to fight. They try to warn strange animals away. Dogs and wolves draw back their upper lips to show their fangs in a warning. A skunk does the same thing, and adds a mean stare.

Members of the cat family warn by raising their hair to make themselves look larger. The ears are laid back and the lips are drawn to show the sharp canine teeth. The tail lashes back and forth. They, too, have a mean stare.

Most lizards open their mouths wide as a warning. The harmless Australian frilled lizard shows the bright red inside of its mouth. It also spreads a large yellow, red and blue frilled collar out on both sides of its head.

If a stranger enters a male herring gull's territory, the gull rushes over and pulls up a bunch of grass with his beak. This means, 'get out or I'll fight.' The stranger either leaves or, if he wants to fight, also pulls up a bunch of grass.

Sound Signals. Sound is a good alarm signal, as it reaches others quickly. A frog gives a hasty croak or chirp as it jumps for the pond.

Birds give alarm calls which often tell the type of danger. When a hen sees a hawk circling overhead, she gives a harsh scream which sends her chicks scurrying for cover. If it is something on the ground, her alarm call is much softer — just a *cut-cut-cut-caaaaaah!* If a blackbird sees danger overhead, it gives a long *seeee!* If the danger is on the ground, the call is just *chuck!*

Alarm sounds aren't always vocal. A rabbit gives a warning thump with its long hind legs. The beaver uses his tail to slap the water before diving under.

Sound signals are also warnings. When threatened, some snakes hiss. A few snakes make a warning rasping sound by rubbing their rough-edged scales together. Rattlesnakes shake

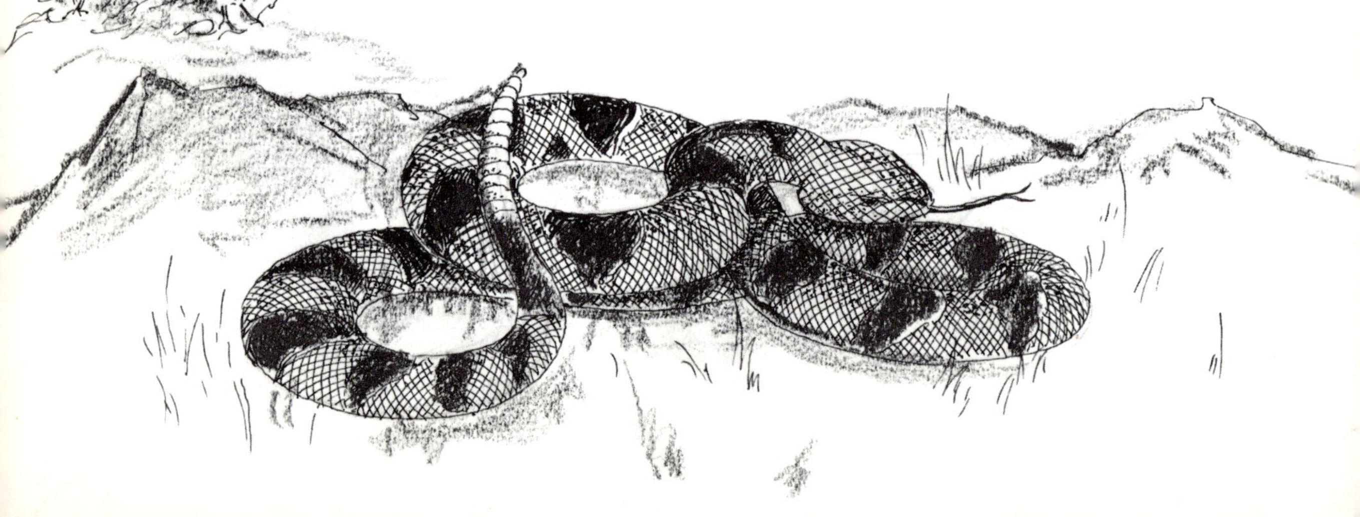

the horny rattles on the end of their tails. The harmless gopher snake pretends to be a rattlesnake by thrashing its tail around in the dry leaves to make a rattling sound.

A porcupine rattles its quills to say, "stay away." Dogs, coyotes and wolves warn with a growl. Members of the cat family hiss or snarl. When two male bears meet, they both begin to cough. The largest bear with the worst cough usually forces the other to leave.

Non-vocal sounds also give a warning. Gorillas beat on their chests with their cupped hands. Chimpanzees make a drumming noise by striking the large roots at the base of a tree with a stick.

Rallying calls gather a flock of birds together to attack an enemy. When a crow spots an owl during the daytime, it gives a harsh call. All crows within hearing hurry to help attack their enemy.

CHAPTER SIX

Studying Animal Communication

Why do we study animal communication?

One of the reasons is to learn more about the ways human beings communicate. In this way we can learn to understand and get along with one another better.

Scientists who study how birds learn to sing are applying their findings to the problems of teaching the deaf to talk.

People who work with animals need to know their communication signals.

Scientists have learned to use recorded alarm or distress signals to drive away flocks of gulls, starlings, crows, and blackbirds that have become pests. In some areas huge flocks of these birds have destroyed crops and caused diseases. Scientists are also using recorded calls to try to drive rats and mice out of buildings and ships.

Animal communication is being used to control insects. Males can be lured into traps by using the scents given off by females. A recording of the whine made by a female mosquito's wings has been used to lure the males into a trap.

Today we are sending people into space. It may be that we will find other beings out there who are unable to speak our languages. How can we communicate with them? The studies being made of ways to exchange signals with living things that cannot talk our languages may help us.

Would you like to study animal communication? You should study many sciences in school. But you can begin your studies by watching the animals you see — especially your own pets. What is your dog trying to tell you and how does it do it? Take notes about what happens. After a time, when you look over your notes, you may begin to get a picture of what your dog is trying to tell you.

The library and your school have books to help you understand the things you see and hear. Don't think you know what an animal is thinking. Animals do not think the way humans do. Be doubtful of everything you see, hear, and read — check everything. When you do all of these, you will be a scientist and a student of animal communication.

Glossary

antennae the organs of touch or hearing on the heads of insects and other animals

chemical signals messages given with particles of odor-producing chemicals. *See also* pheromones

chemical trail a path which can be followed by smell

communication the sending and receiving of signals. Exchange of information

display visual communication by color, shape, or movement, for courtship or defense

dominant animal the ruling animal – often the strongest

eardrum membrane in ear which vibrates as sound waves pass through. Nerve endings in the eardrum then send sound signals to the brain.

grooming stroking or cleaning the fur or feathers of another animal

larynx voice box in certain animals

lower-ranking animals those that follow and obey the dominant animal

mammals warm-blooded animals whose young are born alive and are usually fed with milk from the mother's body

migration a seasonal movement of animals from one place to another

organ a part of the body having a certain function
pheromone chemical substance given off from an animal's body for purposes of communication
retina the part of the eye that receives sight signals (pictures) and sends them to the brain
senses sight, hearing, smell, touch, and taste
smell signals messages received from tiny chemical particles in the air
social animals those living in groups
sound waves vibrations in air or water which are recognized as sound by the brain
species a group of living things having some characteristics in common
syrinx vocal organ of birds
territory the area which an animal or group of animals defends
touch signals messages given and received by touch
vertebrates animals with backbones
visual signals messages received by the eyes or sent in such a manner as to be received by the sight organs of others
vocal cords bands of tissue in the throat used in making sounds
vocal sacs pouches where air can be stored to be used in sound making

Index